Government Accountability Office Bid Protests in Air Force Source Sel~

Evidence and (
Executive Summc

Frank Camm, Mary E. Chenoweth,
John C. Graser, Thomas Light,
Mark A. Lorell, Susan K. Woodward

Prepared for the United States Air Force

PROJECT AIR FORCE

The research described in this report was sponsored by the United States Air Force under Contract FA7014-06-C-0001. Further information may be obtained from the Strategic Planning Division, Directorate of Plans, Hq USAF.

Library of Congress Cataloging-in-Publication Data

Government Accountability Office bid protests in Air Force source selections : evidence and options / Frank Camm ... [et al.].
    p. cm.
  Includes bibliographical references.
  ISBN 978-0-8330-5099-1 (pbk. : alk. paper) — ISBN 978-0-8330-5167-7 (pbk. : alk. paper)
  1. Defense contracts—United States—Evaluation. 2. United States. Air Force—Procurement—Evaluation. 3. Letting of contracts—United States. 4. United States. Government Accountability Office. I. Camm, Frank A., 1949- II. Project Air Force (U.S.) III. Rand Corporation.

  KF855.G678 2012
  346.7302'3—dc23

                                                               2011022736

The RAND Corporation is a nonprofit institution that helps improve policy and decisionmaking through research and analysis. RAND's publications do not necessarily reflect the opinions of its research clients and sponsors.
RAND® is a registered trademark.

Published 2012 by the RAND Corporation
1776 Main Street, P.O. Box 2138, Santa Monica, CA 90407-2138
1200 South Hayes Street, Arlington, VA 22202-5050
4570 Fifth Avenue, Suite 600, Pittsburgh, PA 15213-2665
RAND URL: http://www.rand.org/
To order RAND documents or to obtain additional information, contact
Distribution Services: Telephone: (310) 451-7002;
Fax: (310) 451-6915; Email: order@rand.org

Preface

In 2008, the U.S. Air Force and the Office of the Secretary of Defense (OSD) began a series of intense reviews of Air Force source selection policies and practices. As part of this effort, during the summer of 2008, the Office of the Assistant Secretary of the Air Force for Acquisition (SAF/AQ) asked RAND Project AIR FORCE (PAF) to identify specific changes in policies and practices that could improve Air Force performance in Government Accountability Office (GAO) bid protests.

This document reports the findings of the research project, "Air Force Source Selections: Lessons Learned and Best Practices," which was conducted within the Resource Management Program of PAF in fiscal year (FY) 2009. This project examined the Air Force's recent experience with bid protests before GAO. The Air Force asked PAF to examine all source selections, paying special attention to how the Air Force conducts source selections in large acquisitions.

PAF conducted this analysis at the request of Gen Donald Hoffman, as former military deputy to SAF/AQ; Lt Gen Mark D. Shackelford, SAF/AQ; and Roger S. Correll, then–Deputy Assistant Secretary of the Air Force for Contracting (SAF/AQC). They asked PAF to identify specific changes that the Air Force can make to its source selection policies and processes for complex acquisitions to reduce the rate of successful protests.

To do that, this study used a variety of analytic methods, including a review of relevant government documents, interviews with relevant officials inside and outside the Air Force, econometric analysis of data from administrative federal databases, and a detailed review of the protests associated with two Air Force programs, the Combat Search and Rescue Recovery Vehicle (CSAR-X) program and the Aerial Refueling Tanker Aircraft (KC-X) program. The monograph's findings should interest policymakers and their staffs with responsibility for these programs, for source selection more generally, and for federal policies associated with bid protests in source selections. The econometric methods may also interest those seeking to use administrative databases to test hypotheses about factors that affect policy outcomes that can be described with quantitative data.

The companion documents for this report are:

- *Government Accountability Office Bid Protests in Air Force Source Selections: Evidence and Options,* Frank Camm, Mary E. Chenoweth, John C. Graser, Thomas Light, Mark A. Lorell, Rena Rudavsky, and Peter Anthony Lewis (DB-603-AF). This is the primary PAF research product that this document summarizes.
- *Analysis of Government Accountability Office Bid Protests in Air Force Source Selections over the Past Two Decades,* Thomas Light, Frank Camm, Mary E. Chenoweth, Peter Anthony Lewis, and Rena Rudavsky (TR-883-AF). This provides details on the methods and findings of our statistical analyses of patterns in Air Force experience with GAO bid protests since 1990.

RAND Project AIR FORCE

RAND Project AIR FORCE (PAF), a division of the RAND Corporation, is the U.S. Air Force's federally funded research and development center for studies and analyses. PAF provides the Air Force with independent analyses of policy alternatives affecting the development, employment, combat readiness, and support of current and future air, space, and cyber forces. Research is conducted in four programs: Force Modernization and Employment; Manpower, Personnel, and Training; Resource Management; and Strategy and Doctrine.

Additional information about PAF is available on our website: http://www.rand.org/paf.html

Contents

Preface iii
Summary vii
Acknowledgments xi
Abbreviations xiii

CHAPTER ONE
Introduction 1

CHAPTER TWO
Tracking the Patterns and Trends of Bid Protests 3
The Air Force Should Track Both Corrective Actions and
    GAO-Sustained Protests 3
Overall Air Force Experience with Bid Protests Has Been Positive 4
Significant Trouble Will Persist in a Small Number of Sophisticated
    Protests 5

CHAPTER THREE
Potential Proactive Defenses Against Bid Protests 9
Recognize a Bid Protest as an Adversarial Proceeding with Finely
    Tuned Rules. 9
Simplify and Clarify Selection Criteria and Priorities 13
Focus Formal Cost Estimates on the Instant Contract 16
Tighten Discipline Throughout the Source Selection 21
Tailor Quality Assurance to the Needs of a Less-Experienced Source
    Selection Workforce 25

The Right Kind of External Quality Assurance Can Help 27
New Data Could Help the Air Force Target Its Efforts 28

CHAPTER FOUR
Closing Thoughts 31

Bibliography 35

Summary

When an offeror participating in an Air Force source selection believes that the Air Force has made an error that unjustly "prejudices" its chance of winning the source selection, the offeror can file a protest with GAO. During FY 2000 through FY 2008, the Air Force experienced such bid protests in an average of 93 contract awards a year, and GAO sustained an average of three of these protests a year.

The need to change three source selections a year does not sound like a serious problem in an organization that buys as much as the Air Force does. Unfortunately, a number of the protests that GAO sustained during this period—namely, protests in the CSAR-X and KC-X programs—were highly visible and caused significant disruptions in resource and operational planning in the Air Force. OSD temporarily suspended the Air Force's control of the KC-X source selection. Complications caused by the CSAR-X protest sustainment ultimately helped lead OSD to cancel the program. Consequently, the Air Force and OSD began a series of intense reviews of Air Force source selection policies and practices.

As part of this effort, during the summer of 2008, SAF/AQ asked PAF to identify specific changes in policies and practices that could improve Air Force performance in GAO bid protests. The Air Force asked PAF to pay special attention to how the Air Force conducts source selections in large acquisitions. This executive summary synthesizes findings from a documented briefing that addresses these requests.

Our analysis found that, during the 1990s, the number of unwarranted protests dropped markedly, leaving the Air Force in a position to focus on protests that were more likely to require corrective action. Since 2001, the numbers of corrective actions and of merit reviews per 1,000 contract awards have slowly dropped. The Air Force should be careful to protect the policies and practices that have supported this pattern of steadily improving performance.

The threat manifested in the CSAR-X and KC-X programs appears to be relatively new in character and so does justify significant adjustments in policies and practices in appropriate circumstances. But we expect this threat to present itself in a relatively small number of acquisitions—the large, complex ones, with relatively large stakes for the participants—and it is probably more likely when the participants understand how to pursue sophisticated protests. Such protests will continue and could increase in number until the Air Force demonstrates that it can effectively counter them. The Air Force should focus its countermeasures on the places where the threat is greatest. That should make it easier to tailor the countermeasures to the circumstances and to choose the set of measures best suited to helping the Air Force avoid future costs and delays, such as those associated with the protests sustained in the CSAR-X and KC-X source selections.

In particular, the Air Force can take the following steps to reduce the risks associated with sophisticated protests:

- Recognize a bid protest as an adversarial proceeding with finely tuned rules. Pay more attention to how GAO views a bid protest. Expect GAO to view the priorities associated with running a source selection differently than the Air Force does. Be prepared to pursue the Air Force's interests within the constraints imposed by GAO's priorities.
- Simplify and clarify selection criteria and priorities. The Air Force is already moving aggressively in this direction. The new approach that the Weapon Systems Acquisition Reform Act (WSARA) prescribes for capability and requirements determination could help clarify the relative importance of requirements in ways that promote this goal.
- Focus formal cost estimates on the instant contract. Again, the Air Force is already moving aggressively in this direction. As it

does so, it should be clear in its source selections how and why the cost estimates it uses to discriminate among proposals differ from the cost estimates it uses in its submissions to the Defense Acquisition Board.

- Tighten discipline throughout the source selection. The Air Force plans to rely more heavily on external review processes to enhance discipline. External review can help; greater involvement of attorneys as part of any external review should be especially helpful. But the issues arising in sophisticated protests can ultimately be addressed only by hands-on, close attention to detail that an external review team cannot perform. Tools are available to support discipline and simplify internal review. New forms of training and coaching can also help.
- Gather new data to help the Air Force target its efforts. More attention to the costs imposed by different forms of protest could help the Air Force determine how much to spend to avoid these costs. Better data on the extent to which sustained protests actually change who wins a competition could help the Air Force inform GAO about when an error is actually likely to prejudice any offeror and hence justify a sustainment.

Acknowledgments

This project would not have been feasible without the cooperation, assistance, and goodwill of many people knowledgeable about Air Force source selection and the GAO bid protest process. We give special thanks to Roger S. Correll, Randall Culpepper, and Maj Brett Kayes, within SAF/AQC, who helped us frame the analysis and provided valuable documents and introductions to relevant contacts.[1] Even a cursory examination of this monograph will quickly reveal how much we benefited from the foundational empirical analysis by Maj Brett Kayes. We also thank Sarah Dadson, Olalani Kamakau, Michael J. Maglio, Sharon Mule, and Pamela Schwenke for valuable insights.

Others in the Office of the Secretary of the Air Force gave us important insights into recent Air Force experience with bid protests and the ongoing development of the Acquisition Improvement Plan. They include Lt Col Carole Beverly; Lt Col Jon C. Beverly, Office of the Excutive Action Group of the Assistant Secretary of the Air Force for Acquisition (SAF/AQE); Kathy L. Boockholdt, Office of the Air Force Acquisition Center of Excellence (SAF/ACE); Maj Jon Dibert (SAF/AQXD); Blaise J. Durante, Deputy Assistant Secretary for Acquisition Integration, Office of the Assistant Secretary of the Air Force for Acquisition (SAF/AQX); Patrick M. Hogan, director of SAF/AQXD; James A. Hughes, Jr., Deputy General Counsel for Acquisition, Office of the Air Force General Counsel (SAF/GCQ); Robert Pollock

---

[1]   All offices and ranks are current as of the time of the research.

(SAF/ACPO); and Col Neil S. Whiteman, chief of the Air Force Legal Operations Agency/Commercial Litigation (AFLOA/JAQ).

Michael Golden, at GAO, was generous in sharing his rich understanding of the bid protest process, while carefully skirting any direct discussion of the CSAR-X and KC-X cases highlighted here—cases that GAO still considered to be open when we talked with him.

We spoke at length to government and contractor personnel with direct personal knowledge of the CSAR-X and KC-X source selections and bid protests—from inside and outside the source selections. When we did so, we promised them anonymity to make it easier for them to speak frankly. They were uniformly generous with their time, giving us access to relevant documents and sharing lessons learned that they had developed from their own experiences. We look forward to seeing those lessons learned become public so that all can benefit directly from them. In the meantime, they have given us valuable insights on how to improve Air Force policies in the future.

At RAND, Laura Baldwin oversaw this work and took an active interest in it throughout. Elliott Axelband, Paul Heaton, and Bruce Held provided valuable reviews of the discussions of acquisition policy, economic, and legal issues in the document. Susan Gates helped us gain access to and understand personnel data she had used in the past. Judith Mele similarly helped us interpret contracting data that she had worked with in the past. Our editor, Patricia Bedrosian, significantly improved clarity and internal consistency throughout this and its sibling publications. Megan McKeever was always ready to provide helpful administrative support.

We thank them all but retain full responsibility for the accuracy and objectivity of the analysis presented here.

Abbreviations

AFLOA/JAQ	Air Force Legal Operations Agency/Commercial Litigation
AIP	Acquisition Improvement Plan
ASC	Aeronautical Systems Center
BAA	Buy American Act
CSAR-X	Combat Search and Rescue Recovery Vehicle
DAB	Defense Acquisition Board
DoD	Department of Defense
EMD	engineering and manufacturing development
EN	evaluation notice
FY	fiscal year
GAO	Government Accountability Office
KC-X	Aerial Refueling Tanker Aircraft
MER	Manpower Estimate Report
MPLCC	most probable life-cycle cost
NAVAIR	Naval Air Systems Command
OSD	Office of the Secretary of Defense

PACTS	Protest and Congressional Tracking System
PAF	Project AIR FORCE
RFP	request for proposal
SAF/ACE	Secretary of the Air Force, Acquisition Center of Excellence
SAF/AQ	Assistant Secretary of the Air Force for Acquisition
SAF/AQC	Deputy Assistant Secretary of the Air Force for Contracting
SAF/AQE	Executive Action Group of the Assistant Secretary of the Air Force for Acquisition
SAF/AQX	Assistant Secretary for Acquisition Integration, Office of the Assistant Secretary of the Air Force for Acquisition
SAF/GCQ	Deputy General Counsel for Acquisition, Office of the Air Force General Counsel
USAF	U.S. Air Force
WSARA	Weapon Systems Acquisition Reform Act

CHAPTER ONE

Introduction

When an offeror participating in an Air Force source selection believes the Air Force has made an error that unjustly "prejudices" its chance of winning the source selection, the offeror can file a protest with the Government Accountability Office (GAO).[1] During fiscal year (FY) 2000 through FY 2008, the Air Force experienced such bid protests in an average of 93 contract awards a year, and GAO sustained an average of three of these protests a year.[2]

The need to change three source selections a year does not sound like a serious problem in an organization that buys as much as the Air Force does. Unfortunately, two protests that GAO sustained during this period—namely, protests in the Combat Search and Rescue Recovery Vehicle (CSAR-X) helicopter program and the Aerial Refueling Tanker

---

[1] This protest does not go to the side of GAO best known for generating public policy studies for the Congress. Rather, it goes to the GAO Office of General Counsel, which is administratively separate from the rest of GAO and operates under different governance arrangements.

[2] These numbers are based on data from the Protest and Congressional Tracking System (PACTS) database and counts of root B-numbers for FYs 2000–2008. GAO gives a unique B-number to each protest filing. One protest can have several B-numbers. Each source selection has one "root B-number"—for example, B-299145; each protest filing associated with that source selection has the same root B-number and a different suffix—for example, .2 in B-299145.2. Hence, a count of root B-numbers counts the number of source selections in which protests occur. A count of B-numbers counts the number of protest filings that GAO uses to manage these protests. PACTS is explained in more detail in Thomas Light, Frank Camm, Mary E. Chenoweth, Peter Anthony Lewis, and Rena Rudavsky, *Analysis of Government Accountability Office Bid Protests in Air Force Source Selections over the Past Two Decades,* Santa Monica, Calif.: RAND Corporation, TR-883-AF, 2012.

Aircraft (KC-X) program—were highly visible and caused significant disruptions in resource and operational planning in the Air Force. The Office of the Secretary of Defense (OSD) temporarily suspended the Air Force's control of the KC-X source selection. Complications caused by the CSAR-X sustainment ultimately helped lead OSD to cancel the program. Consequently, the Air Force and OSD began a series of intense reviews of Air Force source selection policies and practices.

As part of this effort, during the summer of 2008, the Office of the Assistant Secretary of the Air Force for Acquisition (SAF/AQ) asked RAND Project AIR FORCE (PAF) to identify specific changes in policies and practices that could improve Air Force performance in GAO bid protests. The Air Force asked PAF to give special attention to how the Air Force conducts source selections in large acquisitions. This executive summary synthesizes findings from a larger report that addresses these requests.[3] It begins with an overview of broad patterns in Air Force experiences with GAO bid protests. These patterns suggest that the Air Force should (1) expand the range of corrective actions that it tracks and (2) focus any efforts to adjust its source selection policies on acquisitions that are likely to attract "sophisticated protesters." The summary describes this emerging type of protest and then reviews a series of adjustments that the Air Force could make to reduce the likelihood that sophisticated protests succeed and to limit the costs that they impose on the Air Force when they do. These suggestions are generally consistent with the Air Force's current direction as defined in the Acquisition Improvement Plan (AIP),[4] but they refine some directions and offer some potential extensions.

---

[3] Frank Camm, Mary E. Chenoweth, John C. Graser, Thomas Light, Mark A. Lorell, Rena Rudavsky, and Peter Anthony Lewis, *Government Accountability Office Bid Protests in Air Force Source Selections: Evidence and Options,* Santa Monica, Calif.: RAND Corporation, DB-603-AF, 2012.

[4] U.S. Air Force, *Acquisition Improvement Plan,* Washington, D.C.: SAF/AQ, May 4, 2009c.

CHAPTER TWO

Tracking the Patterns and Trends of Bid Protests

The Air Force Should Track Both Corrective Actions and GAO-Sustained Protests

When a protest occurs, the Air Force can voluntarily take corrective action to address an error at the beginning of the process. For example, it can offer to reevaluate proposals submitted, give offerors an opportunity to adjust their proposals, change the offerors included in the source selection, rewrite the request for proposal (RFP), and even award the contract to a different winner. GAO normally will dismiss a protest if the agency's corrective action addresses the protest issues. From FY 2000 to FY 2008, the Air Force experienced 836 protests. It offered corrective action in 273 or 33 percent of these protests. It ultimately suffered GAO-sustained protests in only 29 or 3 percent of these.

No empirical data are available to directly compare the costs to the Air Force of corrective actions that the Air Force offers voluntarily and of those that GAO induces the Air Force to provide. Our analysis suggests that the costs of individual GAO-induced corrective actions are likely to be higher than those of individual voluntary corrections, but we cannot say how much higher. That said, the number of corrective actions offered up front (273) is about nine times greater than the number of corrective actions suggested by GAO at the end of a review (29). This pattern tells us that corrective actions offered by the Air Force up front are important, even if the costs of individual actions are less than those for GAO-induced actions. The Air Force and GAO have traditionally tracked Air Force performance by focusing on the outcomes of sustained protests. The Air Force should track both sets of outcomes to understand its own performance with regard to bid protests.

Overall Air Force Experience with Bid Protests Has Been Positive

Looked at broadly, Air Force experience with GAO bid protests over the last two decades represents a positive story. As shown in Figure 2.1, the total number of Air Force protests as a share of the total number of contract awards fell dramatically from FY 1994 through FY 2008—about 65 percent or about 7 percent per year. Sustained protests are so unusual that they hardly register relative to the total number or the value of contract awards. On average, from FY 2000 through FY 2008, GAO sustained one protest for every $20 billion the Air Force spent in acquisitions. So few sustained protests have occurred that it is impossible to discern any trend in them. On the other hand, the Air Force has offered corrective actions in notice-

Figure 2.1
Trends in Bid Protests and Corrective Actions, 1994–2008

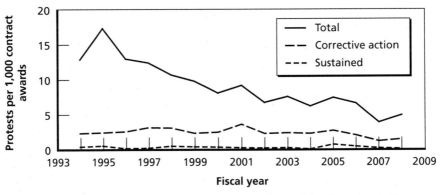

SOURCE: RAND analysis of data from administrative federal databases (PACTS) and the Federal Procurement Data System—Next Generation.

NOTE: The chart normalizes the number of protests and corrective actions that have occurred each year by stating them relative to thousands of contract awards.

AND *MG1077-1*

able numbers. The number of corrective changes per contract award bobbed around until FY 2001 and then began a long downward trend. Through the 1990s, the Air Force offered corrective actions in about 0.3 percent of contract awards. From FY 2001 to FY 2008, the percentage has fallen fairly steadily, ending well under 0.2 percent of contract awards. All of these trends point to steady improvement over time.

The Air Force has had a higher success rate in defeating individual grounds for protest than the standard performance statistics tracked can detect. No one maintains an official record of how many discrete grounds for protest occur within any one acquisition, but Air Force officials familiar with the CSAR-X source selection agree that the first effort to overturn the CSAR-X source selection decision involved about 130 separate issues. GAO rejected the protest grounds for all of these but one. The effort to overturn the KC-X source selection decision also involved about 130 issues; GAO rejected the protest grounds for all of these but eight. GAO only had to sustain one issue in a protest to force a change in the Air Force approach, but it is worth keeping in mind how many faulty claims protesters made to achieve a sustainment on a single issue.

Recent historical experience suggests that, looked at broadly, Air Force source selection policies and practices work. Improvements are always possible, but any sense of crisis associated with recent events should be kept in perspective. The broad record suggests that far-reaching changes in Air Force source selection policies and practices are not required and, in fact, might even lead to more harm than good. The Air Force should target change where the problem is largest.

Significant Trouble Will Persist in a Small Number of Sophisticated Protests

Despite these positive trends, the Air Force continues to experience serious protests in large, complex acquisitions that present high stakes to their participants. The CSAR-X and KC-X programs are examples of such acquisitions, and their source selections provide insights that can help guide efforts to reduce the likelihood that such protests will succeed in the future. However, these acquisitions are not entirely representative of the acquisitions where we expect continuing trouble. Understanding how they were unusual relative to other large Air Force acquisitions will prevent the Air Force from making unnecessary policy adjustments.

First, both acquisitions use preexisting designs with characteristics that are relatively well known to all potential offerors: The CSAR-X design was expected to draw heavily on the design of preexisting military and commercial helicopters, and the KC-X design was expected to draw heavily on a preexisting commercial aircraft design. Using preexisting designs makes it difficult to define functional and performance requirements clearly while preserving competition: The more precisely requirements are stated, the easier it would be for potential offerors to compare available designs against the requirements and determine which design would likely win. Consequently, it would not be rational for anyone but the apparent front-runner to enter the competition. Knowing this, the Air Force framed its requirements with exceptional subtlety to induce enough offerors to participate in each of these source selections. Lack of clarity about requirements ultimately provided one of the grounds for the sustained protest in the KC-X program. The Air Force should not expect this problem to be as serious in circumstances where the attributes of potential proposed designs are not so well documented in the public record. However, the Weapon Systems Acquistion Reform Act (WSARA) is likely to induce a situation like this to arise more often in major system developments.[1]

Second, the CSAR-X and KC-X acquisitions are unusual in that both involve foreign offerors. The Air Force was clear throughout that it had no position regarding whether an American or a foreign offeror

---

[1] WSARA (Public Law 111-23, May 22, 2009) requires that a number of actions, such as competitive prototyping and a preliminary design review, occur before the Air Force selects a source to complete development of a new system through engineering and manufacturing development (EMD). Such actions are designed to mature systems offered for EMD. Such maturation in all likelihood will increase public understanding of the systems being offered, complicating the development of requirements that can sustain competition during a source selection in the same way that public knowledge of capabilities in the CSAR-X and KC-X source selection complicated those source selections.

prevailed so long as the award complied with the Buy American Act.[2] But the potential of a foreign winner increased congressional interest in both competitions and created a heated political environment that heightened the stakes of these competitions. We found no evidence that congressional interest affected decisionmaking within the Air Force or GAO in either source selection. However, high political interest, at a minimum, complicated decisionmaking by drawing more pointed external interest than usual to these acquisitions.

Although the Air Force should be cautious about generalizing from CSAR-X and KC-X source selections, these acquisitions share two other characteristics that we do expect to see repeatedly in a small number of Air Force acquisitions in the future. First, each acquisition had a high dollar value. In the KC-X, this value was magnified by the expectation of two follow-on acquisitions of approximately equal value; success in the first would likely support further success in the follow-ons. The net revenue associated with large acquisitions is likely to encourage any loser in a source selection to launch an aggressive protest, even if the cost of that protest is substantial. Second, the winner of each competition was likely to emerge as the global front-runner for the capability in question, gaining an advantage not only in the U.S. market, but in markets for other nations seeking the same capability. Technology maturation and learning in production would give the winner an advantage in future competitions, which any challenger would have difficulty overcoming. In the extreme, a failure in either source selection could foreclose a loser's future business opportunities in a large share of its potential markets.

In the face of these considerations, the chief executive officer of any losing offeror would have a fiduciary obligation to pursue a protest, even if the chances of success were small. The stakes associated with each of these source selections were so high that even a costly campaign to protest the decision could be not only worthwhile from the loser's

---

[2] The Buy American Act (BAA—41 U.S.C. § 10a–10d, 1933) exempts contractors in "qualifying countries" from the BAA, treating them the same as U.S. companies for contract award purposes. U.S. firms are treated (reciprocally) as domestic contractors in procurements that occur in qualifying countries.

point of view but absolutely necessary to satisfy stockholders. These stakes make it easy to appreciate the intensity and sophistication of the protest efforts launched in both these source selections. It is hard to imagine that any traditional government jawboning could have dissuaded the losers from pursuing these efforts.

The main lesson that the Air Force can learn from the pattern of protests that we have described is that the sophisticated nature of the CSAR-X and KC-X protests is likely the product of circumstances that the Air Force will face again. Sophisticated protesters appear to be learning how to achieve GAO sustainments, and their outside counsels are becoming increasingly capable of supporting or even designing such campaigns. The waves of grounds for protest that both the CSAR-X and KC-X programs experienced show clear marks of an orchestrated process to generate grounds for protest designed in part to release significant numbers of government documents that would likely provide material for still more grounds for protest. One of the lessons these protesters are learning is that it is hard for the Air Force to surge its resources to counter sophisticated protests. Because protesters can surge protest resources more effectively than the Air Force, they may see a growing opportunity in large, sophisticated protests designed to stretch Air Force capabilities. As their outside counsels learn how to do this, the counsels can be expected to offer this capability to other companies in competitions involving smaller stakes.

In the face of this threat from sophisticated protesters, the Air Force has a simple and direct defense: Learn enough about how these campaigns work so that it can limit their success, either by foreclosing opportunities for protest during the design and execution of a source selection or by being prepared for the protests when they come. A failure to build such a defense will likely encourage still more sophisticated protests and more sustainments against the Air Force's interests.

CHAPTER THREE

Potential Proactive Defenses Against Bid Protests

The Air Force should focus on changing source selection policies and practices in ways that help it counter such sophisticated protests. As it does so, it should be cautious not to make changes that could hurt its long-term success in dealing with the vast majority of GAO bid protests. It should also be cautious about focusing too much on issues specific to the CSAR-X and KC-X protests. Rather, it should seek proactive defense against sophisticated protests in large, complex acquisitions that present particularly high stakes to the participants. The Air Force AIP has already begun to implement changes that move the Air Force in the direction we suggest here. But several opportunities exist to refine the AIP or move beyond it.

Recognize a Bid Protest as an Adversarial Proceeding with Finely Tuned Rules

Air Force personnel are often puzzled by GAO's arm's-length stance during bid protests. To them, GAO's approach to bid protests seems technical, counterintuitive, and not supportive of the military mission that most Air Force personnel take for granted. Just beneath the surface, one senses a certain suspicion among them that the Air Force and GAO are not on the same team. Viewed appropriately, that suspicion is correct. GAO and the Air Force have separate and distinct missions that naturally come into conflict in bid protests. Air Force personnel will be more successful in bid protests when they understand better how GAO views them.

Congress gave the responsibility for bid protests to an independent third party—GAO—to ensure that the principles of acquisition regulations were given adequate attention, quite separate from the military missions of the services in which acquisition occurred. Consequently, GAO views itself as an impersonal arbiter, simply applying the statutory guidance that defines its mission to the specific facts of individual bid protests. When ambiguities arise during source selection (e.g., about such matters as the thresholds for performance requirements), Air Force personnel often find it natural and reasonable to resolve disagreements in ways that favor the clear priorities of the warfighter. GAO, on the other hand, interprets the language in any particular source selection without falling back on military priorities when the language is unclear.

That is not to say that GAO will not defer to military judgment. In fact, in the majority of merit reviews, GAO shows great deference to military judgment where the source selection documents clearly define the role for such judgment. When the language in these documents speaks clearly for itself, GAO accepts this language on its own terms. When the language does not speak clearly, GAO does not resort to the military reasoning that might have resolved any uncertainty in a purely Air Force setting.

Other than the procedural rules for the conduct of bid protests, GAO does not promulgate procurement-related regulations.[1] But its published bid protest decisions provide a comprehensive body of law that the Air Force can use to (1) design and execute source selections to avoid protests and (2) make better-informed decisions about when to offer corrective action when protests occur to avoid GAO review. When GAO explains its decision in one bid protest, it relies heavily on decisions it has made in similar cases in the past to explain its reasoning. The record of GAO decisions provides a public statement of how GAO officials think. The better the Air Force can anticipate how GAO will treat a new set of facts, the better able the Air Force will be to avoid protests in the first place and respond to them when they occur.

---

[1] For example, it holds no formal rulemaking procedures under the auspices of administrative law to define formal regulatory rules that it then promulgates and enforces.

Given the basically legal nature of the bid protest process as it exists today and is likely to continue to be in the future, it is natural to imagine an expanded role for Air Force attorneys and potentially for third-party attorneys. Attorneys working as part of a government "red" team with personnel with other contracting and acquisition skills could proactively scour decisions and documents generated during Air Force source selections with the object of anticipating how a protester's attorneys would do the same after a protest starts.[2] This review would not be a standard review designed to ensure orderly program execution. Rather, it would deliberately seek weak points, expecting that protesters will use any weak points they discover to leverage their resources to greatest effect. That is, this type of review would anticipate that, in source selections likely to involve sophisticated protesters, the endgame will occur in an adversary process in which the Air Force will be asked to defend all of its actions through the course of the source selection.

Adding attorneys and steps to the source selection process would almost certainly raise the administrative costs of the source selection and add calendar time to its schedule. Additional personnel cost money to train and sustain; this expense will be easiest to justify if these personnel can reduce other costs still more. Similarly, because the tasks that red teams perform will almost surely be on the critical path of any source selection, it will be easiest for the Air Force to justify their presence there if they can shorten the calendar time associated with other steps on the critical path—for example, the calendar time required to complete evaluation or deal with a sustained protest.[3] The protest-

---

[2]   Such a team might also productively include engineers, logisticians, contracting specialists, cost analysts, and so on. The Air Force currently uses review teams with such personnel. The red team we describe would not replace these teams. It could act as an integral part of existing review teams.

[3]   In a complex activity with many interrelated steps, the *critical path* identifies a set of steps that, taken together, define the minimum amount of calendar time required to complete all steps. Steps on the critical path cannot be rearranged or rescheduled to reduce the calendar time required to complete the activity. Steps associated with activity-wide review often lie on the critical path, because activity-wide status must be assessed in these steps before any individual part of the activity can move forward. The legal reviews contemplated here lie within such steps in a source selection. Such reviews can shorten the schedule of a source selection only if they reduce the time to complete other steps in the future that also lie on the critical

induced monetary costs and delays experienced in the CSAR-X and KC-X source selections were large relative to the added cost and delay one might hypothetically associate with any red team. But when determining how large a red team should be or how many calendar days to allow for its review, the Air Force should recognize that as more and more person-days are added, the incremental benefit of each additional legal person-day will ultimately fall to a level that is lower than the cost the Air Force expects to pay for such a person-day.

This red-teaming capability need not be wholly organic. The Air Force legal community has been reluctant to use contract attorneys to perform legal tasks that could easily be regarded as inherently governmental.[4] The red team envisioned here involves no activities that could be construed as inherently governmental. It is strictly an advisory activity with no authority to make policy, dispose of government resources, or direct government personnel. In fact, external attorneys might be more effective in this role, because they would not likely assume the traditional roles of government personnel working within a source selection and, if drawn from the same third-party bar that advises sophisticated protesters, might be exceptionally knowledgeable about how these protesters operate.[5] More generally, as long as the Air Force has qualified organic personnel to oversee the work of such third-party red teams, their availability would significantly enhance the Air Force's ability to surge program-specific resources during a source selection to respond to the surge of workload that protesters seek to generate during a source selection and, over the longer term, potentially deter the efforts of sophisticated protesters to overwhelm government

---

path—for example, future reviews within the source selection or by an outside party such as GAO.

[4]   The Federal Acquisition Regulation defines the formal meaning of an "inherently government activity" in Subpart 7.5. The Office of Management and Budget is currently revising its policy on inherently governmental activities and the role of contractors in the government. Changes under way could strengthen the position of those who oppose the use of third-party attorneys within Air Force activities, particularly acquisition activities.

[5]   A similar argument applies to the retired military and civilian personnel the Air Force might consider turning to when staffing red teams. This is a case where knowledge of traditional Air Force processes is not the most desired trait in a contractor.

offices during source selection. This surge capacity could help offset what appears to be one of the major motivations for the increase the Air Force has seen in sophisticated protests in recent years.

Use of third-party attorneys is not a perfect solution. Conflict of interest issues would have to be carefully addressed to separate attorneys supporting the government from those supporting protesters. The government should also be cognizant that, despite any conflict of interest agreement, work for the government today could teach an attorney or law firm how better to support a protester in the future. The Air Force might want to explore appropriate terms of engagement before bringing private-sector attorneys even closer to the Air Force source selection process.

Simplify and Clarify Selection Criteria and Priorities

GAO's most common ground for sustaining a protest is a mismatch between the criteria stated in the request for proposal and the evaluation of these criteria later in the source selection.[6] This accounted directly for three of the ten grounds for the protests sustained in the CSAR-X and KC-X source selections.[7] The simpler and the clearer the criteria and priorities among criteria presented in the request for proposal (1) the less ambiguous these will be when GAO reviews them and (2) the easier it will be for evaluators to execute and document their evaluation in a way that complies with GAO's expectations. Ambiguity in the CSAR-X and KC-X RFPs gave GAO large openings to impose its own judgment about what a "reasonable" person would think the Air Force intended in its request for proposal.

The Acquisition Improvement Plan is moving the Air Force in the right direction by simplifying and clarifying the trade-off process

---

[6] Well before any source selection starts, the Department of Defense (DoD) capability development process identifies requirements for the system to be acquired via the source selection. Acquisition specialists then translate the requirements for this system into selection criteria that evaluators will use during the source selection to assess each proposal.

[7] GAO sustained two protests in the CSAR-X source selection and eight in the KC-X source selection.

that the Air Force uses to weigh the relative importance of various elements of performance and price in an offer and then to determine which among several offers provides the best value to the Air Force.[8] Source selections that use a trade-off process can set a *threshold* and *objective* level for each requirement.[9] The range between the threshold and objective for each requirement constitutes a trade space within which the Air Force can explore options with offerors and negotiate with them to move from a threshold toward an objective in the most cost-effective way possible. To use this method of source selection to greatest advantage, the Air Force requests for proposal must identify thresholds and objectives clearly and explain clearly how Air Force evaluators will assess the value of increases in offered performance across all requirements in the trade space. Advocates believe that this approach has helped the Air Force increase what it gets for the money it ultimately pays the offerors that win source selections.

As advantageous as this approach may be, it probably makes it easier for protesters to have protests sustained. Priorities within the trade space are inherently complex and ambiguous. For example, if the Air Force requires some threshold level of performance for defensive systems on a tanker but rewards performance above the threshold, it can be difficult to define what performance above the threshold is worth to the Air Force until the Air Force sees the specific mechanisms that a proposal offers. At some point, professional, but necessarily subjective, military judgment may be required to assess the contribution of any specific, proposed mechanism. The more complex and ambiguous priorities are, the more opportunities protesters have to demonstrate that (1) an evaluation deviates from the RFP criteria used in a source selection, (2) the Air Force has not treated offerors equally, or (3) the Air Force has used unreasonable methods or judgments to evaluate criteria in the trade space.

---

[8] U.S. Air Force, *Source Selection,* Mandatory Procedure MP5315.3, Washington, D.C.: SAF/AQCP, March 2009b, defines this approach in detail.

[9] If the objective equals the threshold for a requirement, the requirement is not part of the trade space. When this occurs, an offeror must achieve the threshold to be considered responsive. For such a requirement, the Air Force provides no extra credit for any performance beyond the threshold.

The draft RFP for the new round of the KC-X source selection reflects changes in response to AIP guidance that should help future source selections avoid some of the complications this program (and the CSAR-X program) initially encountered. For example, it has dramatically reduced the number of factors included in the trade space for the trade-off among requirements. It has stated the relative importance of different requirements far more clearly. And, in choosing which requirements to give the most attention to in the evaluation, it has considered not only which factors matter most to the Air Force but also which are most likely to discriminate among proposals in the second round.

The importance of being able to use the trade space to discriminate among proposals should lead a source selection to use a smaller trade space than that addressed in WSARA. WSARA encourages the Air Force to consider much broader trade-offs among performance, cost, and schedule while the Air Force defines the capabilities it will seek to achieve by initiating a new program. The assessment of such trade-offs during early capability and development planning—that is, before a source selection ever begins—should resolve many issues that no longer need be addressed during source selection. Resolution could result in many specific requirements that any proposal offered will have to meet in a source selection. Source selection addresses only trade offs that remain unresolved *and that are likely to discriminate among proposals.*

It is easy to confuse the goals of trade-offs addressed before a source selection and the goals associated, under WSARA, with the likely much narrower range of trade-offs addressed during a source selection. The Air Force can help avoid such confusion by being clear in an RFP about how the criteria included in the RFP reflect the requirements that emerge from the resolution of trade-offs during the requirements determination process. If WSARA has the effects desired, such early consideration of trade-offs should help Air Force source selections define the relative importance of the criteria they use and hence explain more precisely how they will address comparisons within the trade space that remains during a source selection itself.

Focus Formal Cost Estimates on the Instant Contract

Explicit formulas for estimating life-cycle costs have invited close questioning by GAO about the reasonableness of Air Force assumptions about the future. Three of the ten grounds for protest that GAO sustained in the CSAR-X and KC-X source selections involved GAO's assessment of Air Force life-cycle cost estimation practices.[10] The Air Force has a strong interest in understanding the likely life-cycle costs of any new system, but the only costs relevant to a source selection are those that will help the Air Force discriminate among proposals. For example, if the Air Force expects the cost of its own organic support of a new system to be the same, no matter which proposal it chooses, that cost is not relevant to the source selection. Where it does expect life-cycle costs to differ among proposals, the Air Force's uncertainties about the future complicate the assessment of these differences. Being precise about an inherently ill-defined future can create opportunities for GAO to object that the Air Force's treatment of future costs is unreasonable. For example, the cost of fuel required to operate tankers is important to the Air Force. What cost of fuel should the Air Force posit over the expected 30-year life of a tanker? No objective, analytic mechanisms exist to resolve uncertainty about this cost.

The AIP has already changed policy to require high-level approval before a source selection uses most probable life-cycle cost as a criterion.[11] Going forward, when the Air Force expects operating and support costs to differ among proposals over the lifetimes of the systems in question, it will follow the lead of Naval Air Systems Command (NAVAIR) and focus cost estimates in a source selection on the "instant contract"—in effect, on the price that a proposal offers for the deliver-

---

[10] One of these, the first sustainment in the CSAR-X program, ultimately turned on a different issue—the Air Force's failure to execute the approach to life-cycle costing implied by its request for proposal. But GAO's discussion of this error paid close attention to the Air Force's approach to life-cycle cost estimation and suggested that a reasonable person would have expected the Air Force to take a different approach than the one it used in evaluation. The two sustainments from the KC-X program explicitly reject how the Air Force estimated specific elements of life-cycle cost, calling them "unreasonable."

[11] U.S. Air Force, 2009c.

ables identified in the request for proposal. Over the period FY 1986 to late FY 2008, NAVAIR had only one protest sustained by GAO.[12] The NAVAIR approach captures engineering information about the future supportability of a system (mean time between failures, mean time to repair, system availability or maintainability, and the like) and evaluates this information as part of the system's technical evaluation. This approach would allow the Air Force to infer likely life-cycle cost implications of different proposals without raising potentially problematic issues about how it views the uncertain future environment in which systems will operate. GAO has accepted this approach in NAVAIR source selections for major systems.

If the Air Force really cares about life-cycle costs, why should it use engineering proxies for future costs?[13] The completeness and quantitative nature of life-cycle cost accounts require stark statements about inherently subjective assumptions regarding a highly uncertain future. By contrast, a supportability subfactor can weigh engineering inputs about all aspects of future system support without specifying in advance precisely how the Air Force will combine information about different engineering inputs to yield a final evaluation for the subfactor. Put another way, subjective military judgment appears to be better than precise cost accounts as a way to address factors about which

---

[12] Brett N. Kayes (Capt, USAF), "Air Force GAO Protest Trend Analysis," briefing, Chart 38, Washington, D.C. (SAF/AQCK), updated September 19, 2008. This sustainment involved an issue unrelated to cost estimation (see U.S. Government Accountability Office, Decision, Matter of Delex Systems, Inc., File B-400403, October 8, 2008).

[13] An irony of the CSAR-X source selection is that GAO misread Air Force requests for information on supportability as evidence that the Air Force cared significantly about differences among proposals in future support costs. In the CSAR-X source selection, Air Force interest in supportability stemmed from concerns that the legacy system that a CSAR-X system would replace could not service new mission requests quickly and reliably enough because its system availability rate was unacceptably low. The Air Force wanted a CSAR-X system to display high availability to reduce the time between receipt of a call for support and execution of a support mission, even if higher availability involved higher future costs. Because the request for proposal did not indicate how the Air Force planned to use data on supportability, GAO misread the Air Force's intent in the source selection. All of this should remind us that, even though supportability can serve as a proxy for life-cycle cost, this is not the only way it can be used. To avoid misunderstanding, the Air Force should be as clear as possible about how it plans to use inputs on supportability or any other factor.

great uncertainty persists. The precision required by cost estimates may create a false sense of certainty about the future that can invite questions about how reasonable it is for the Air Force to use any exact assumption offered. Without asking whether, objectively thinking, it is better to use precise life-cycle cost estimates or subjective military judgment to evaluate statements about the future, it appears reasonable to suggest that GAO decisions in bid protest reviews display this kind of reasoning.

As the Air Force moves toward a more subjective, qualitative approach to evaluating the future supportability of new systems in source selection, it should keep in mind that OSD requires that the Air Force estimate the life-cycle cost of a new system as part of the Defense Acquisition Board (DAB) program review process. The DAB process approaches its Milestone B at almost exactly the same time that the corresponding source selection approaches a final decision. That is, in the future, the Air Force will use one estimate of cost in its source selection process and, nearly simultaneously, a different estimate of cost in its program review process. When it does this, it must ensure that this difference does not confuse GAO.

In the past, GAO did not see the DAB program review estimate, which was broadly understood to be pre-decisional and not suitable for release beyond DoD. Nonetheless, in several recent source selections, GAO has obtained access to this estimate in camera and used it to help assess the reasonableness of the cost estimates used in a source selection. This is likely to happen again in the future. When it does, future Air Force requests for proposal should clearly explain that different definitions of cost are appropriate for the two different decision processes and explain why the source selection uses one while the program review uses another. This will prevent some future GAO official from ruling that the cost estimate in a source selection is "unreasonable" because it differs from the estimate the Air Force uses in an entirely separate but closely related decision process.[14]

---

[14] For example, the difference in the costs relevant to system program review and source selection appears to have confused Senator John McCain, who asked Secretary of Defense Robert Gates, "will the [KC-X] source-selection authority assess most probable life-cycle

The costs relevant to the instant contract in the source selection are a subset of the total costs relevant to the program review, so the only additional effort required to maintain two separate cost estimates is to be clear about which life-cycle costs to associate with the instant contract. To the extent that the Air Force uses a work breakdown structure to build cost estimates from the bottom up, costs relevant to the instant contract should occupy a distinct portion of the work breakdown structure and so be easy to identify. Using the work breakdown structure to distinguish the two cost estimates, then, should help explain the differences between two separate estimates and help clarify for GAO why the two estimates differ.

GAO's assessments of Air Force cost estimation appear to expand its focus from assessing compliance with procedures to more substantive issues. The assessments in recent sustainments basically say that GAO does not approve of how the Air Force uses formal analysis to support its decisions. GAO justified its sustainments on cost issues in the KC-X source selection not by reference to precedents set by earlier GAO bid protest decisions but by reference to a technical handbook on cost estimation.[15] Such a justification has been quite exceptional in the recent GAO bid protest decisions we examined in the course of this project. The Air Force decision to move away from quantifying future costs in source selection should limit its exposure to such GAO judgments, but the evaluation of costs associated with the instant contract will continue to raise similar issues. What can the Air Force learn from GAO's discussions?

First, GAO's discussions of cost estimation in the three grounds for protest sustainment that turned on cost issues were well informed. GAO applied sound costing principles to produce a better approach to cost estimation than the Air Force had used.

---

cost (MPLCC)?" A full assessment of MPLCC is not required to discriminate between the two offerors for KC-X, because many costs do not depend on which system the Air Force chooses. (Senator John McCain, Letter to the Honorable Robert M. Gates, U.S. Senate Committee on Armed Services, Washington, D.C., October 29, 2009.)

[15] U.S. Government Accountability Office, *Cost Assessment Guide: Best Practices for Estimating and Managing Program Costs*, GAO-07-1134SP, Washington, D.C., July 2007c.

For example, in the CSAR-X program, GAO questioned why, if the Air Force planned to adjust its Manpower Estimate Report (MER) in response to early experience with any new CSAR system, it was appropriate to use a MER for a legacy system that in all likelihood would have support requirements that differed from those of new systems. The Air Force has used a MER for decades to estimate the support costs of new systems in program reviews and source selections. When, in response to GAO's first sustained protest in the CSAR-X program, the Air Force clarified that that is precisely what it intended to do—to estimate most probable life-cycle cost of new systems based on the manning levels defined by a MER for a legacy system—GAO backed off and accepted this clarification. But GAO's initial observation was correct—the Air Force's approach, clearly stated in the amended request for proposal that it crafted in response to GAO's first sustainment, was not the approach a reasonable outside observer would expect the Air Force to use in a source selection. At a minimum, because it provided no ability to discriminate between the proposals, it was clear that this approach added no value in the source selection. Although GAO acceded to this flawed approach to costing when the Air Force clearly explained it, the Air Force would benefit from learning from GAO's initial reservations. Doing so would improve Air Force decisionmaking in future source selections—and perhaps in other decision processes that use a similar approach to cost estimation.

GAO was more intrusive in the KC-X source selection. It identified two cases where the Air Force used a flawed approach to estimating future costs. This time, it ruled that these approaches were unacceptable, whether or not the Air Force explained them clearly in the request for proposal. That is, in each case, GAO sustained a protest not because of failure to match evaluation to the plan identified in the request for proposal but because the plan itself was flawed.

To understand GAO's view, consider the second protest GAO sustained on these grounds in the KC-X case. GAO rejected Air Force use of historical data on growth in total development costs in many DoD programs to estimate future growth in one component of development costs—nonrecurring engineering costs—in one program. Why? First, total development costs include many costs other than nonrecurring costs. Comparing the growth that Boeing would likely experience in its nonrecurring engineering costs with the historical growth observed in total development costs in many DoD development programs was equivalent to comparing apples and oranges. The two classes of costs are simply qualitatively different. Second, even if the classes of cost had been the same, the Air Force offered no evidence that Boeing would have the same cost experience that other contractors had had in the past and, in fact, overlooked some potentially contrary evidence available to the Air Force within the source selection itself. As a result, GAO judged the Air Force's assessment of Boeing's likely nonrecurring engineering costs as unreasonable.

Once again, as intrusive as GAO was in this case, its observation was correct. By learning from this GAO judgment, the Air Force could improve decisionmaking in future source selections and elsewhere. The lesson is not just about future cost growth, which the Air Force's new approach to cost estimation in source selection should render irrelevant, but about justifying the historical data used to assess cost estimates that will continue to appear in future source selections.

The good news is that, if GAO continues to take an intrusive stance on cost methodology—and now it has precedent it can cite to do so—there is a good chance that it will use sound cost estimation principles as a basis for such intrusion. That will make it easier for Air Force cost analysts to prepare for future GAO attention to their work in source selections.

Tighten Discipline Throughout the Source Selection

The nature of GAO's implicit rules means that small errors during a source selection can have large consequences. Small errors directly produced three of the ten grounds for protest in the CSAR-X and KC-X source selections. These errors created ambiguities that GAO concluded were significant enough to potentially affect the outcome of the source selection. The Air Force could have avoided these errors only by maintaining close and tight control over the whole source selection to ensure that all parts of the request for proposal were internally consistent and that every one of hundreds of evaluation issues opened during the source selection were properly closed and disposed of.

A quick review of the error that provided the third ground for sustainment that GAO listed for the KC-X source selection illustrates why such errors are so hard to catch. The Air Force told Northrop Grumman twice during discussions that its "initially identified maximum operational airspeed . . . would not be sufficient under current Air Force overrun procedures to achieve required overrun speeds . . . for various fighter aircraft."[16] If the Northrop Grumman proposal could not achieve this requirement, it would be technically unacceptable. The Air Force ultimately accepted Northrop Grumman's proposed solution as satisfying this key performance parameter threshold but never documented the basis for its acceptance. Boeing complained that the Northrop Grumman solution did not appear to meet the threshold requirement. During the protest proceedings, the Air Force could not rebut this complaint.

GAO's decision discusses this protest at considerable length and ultimately concludes that very serious doubts exist about whether Northrop Grumman can meet the threshold. As a starting point for our own analysis, we take as given the Air Force's argument that Northrop Grumman could meet the threshold. Whether Northrop Grumman could or not does not affect our argument. Even if the Air Force was correct in its evaluation, it erred in its documentation of that evaluation and its defense of it in GAO hearings on the question.[17] The more basic error was the failure to close out the evaluation notice that the Air Force opened when it first queried Northrop Grumman about this issue. The KC-X program office used a standard information management system, EZ Source, to track the status of all evaluation notices. Despite the presence of that system, the program office failed to see that it had never explained, in the record, the basis for its judgment

---

[16]   U.S. Government Accountability Office, Decision, Matter of the Boeing Company, Files B-311344, B-311344.3, B-311344.4, B-311344.6, B-311344.7, B-311344.8, B-311344.10, B-311344.11, Washington, D.C., June 18, 2008.

[17]   If its judgment on this issue was not correct, the Air Force faces a more serious problem than those we focus on here. That is a basic substantive failure that none of the solutions we offer here can address.

that the issue was resolved. The KC-X source selection generated hundreds of evaluation notices, so it is not an easy task to track all of them and verify that the evaluation team has documented its resolution of each issue in a substantively sound way—a way that GAO would view as "reasonable." The failure to track this one evaluation notice opened the door for GAO that ultimately led to a sustained protest.

An examination of the CSAR-X and KC-X programs suggests that three factors contributed to such sustainments. Both programs had shortages of properly trained personnel with enough prior experience in complex source selections to prepare them for handling the tough, day-to-day task of ensuring that a request for proposal is aligned with the users' priorities and that the source selection pursues these priorities in an unambiguous, internally consistent and legal way, documenting and controlling evaluation material carefully enough to prevent adversarial attorneys representing protesters from finding residual problems in the record. Ongoing efforts to rebuild the organic Air Force acquisition workforce should help address this problem, but these efforts will take many years to make a significant difference in future Air Force source selections.

Both source selections moved too fast. The KC-X program office felt pressure from above to move forward, at least in part to offset the considerable delays that the program had suffered before this source selection. The CSAR-X program office put a premium on speed, believing that the faster the source selection occurred, the fewer external surprises could occur that could change the basic parameters of the source selection and so add administrative burden and complexity to the process.[18] But combining speed with a shortage of personnel ultimately burned out the key personnel in the source selection. The recent shift in Air Force policy to make source selections event-oriented rather than schedule-oriented could potentially ameliorate some of these pressures

---

[18] The office staff envisioned themselves operating within a classic observation-orientation-decision-action loop in which they wanted to keep their decision cycle as short as possible relative to those in other processes that could affect the source selection. John R. Boyd, "Discourse on Winning and Losing," briefing, 1976.

if the Air Force gives source selections the time required to complete appropriate tasks.

Two types of tools could have helped structure activities in these source selections in ways that made it easier to catch small errors such as those discussed above. One tool would track each evaluation activity described in Section M to the specific requirement that this evaluation addresses and to each piece of information requested in Section L in support of this evaluation activity.[19] Such a tracking tool would facilitate consistency checks across the entire request for proposal and help ensure that the request for proposal describes the specific evaluation methods the Air Force will use to explain how the methods inform requirements and how they use information. Such a mapping, in itself, could clarify the Air Force's plans for evaluation in ways that could (1) make it harder for offerors to claim a misunderstanding about the evaluation and (2) provide a clear benchmark against which to track evaluation. Many Air Force source selections have used a simple matrix that supports such tracking. The draft request for proposal in the new round of the KC-X source selection uses one. Neither the CSAR-X nor the original KC-X source selection used one.

A second tool could track evaluation notices (ENs) from their inception to their final resolution. An effective system of this kind would support ongoing reviews of such questions as the following: (1) Has each question raised in an EN been resolved? (2) When differences of opinion arose within the evaluation team in response to ENs, were these differences decisively addressed and disposed of? (3) Does the record contain sufficient documentation to justify resolution of the question? (4) Does the record include any information not required to track evaluation and justify final decisions? If so, why is it there? Air Force programs often rely on EZ Source, a standard information management system, to regularly check the status of documentation against GAO-informed standards. Both the CSAR-X and KC-X pro-

---

[19] Section L of a request for proposal identifies all the information that an offeror must provide to be responsive to the request for proposal. Section M describes the criteria that the source selection will use to assess each proposal and how that assessment will occur. See Federal Acquisition Regulation, Subpart 14.2, "Solicitation of Bids."

grams used this system. Unfortunately, EZ Source is a cumbersome system that records actions but does not ensure appropriate execution of these actions as much as it could. In the lessons learned exercise conducted at the close of the CSAR-X program, for example, EZ Source was the dominant object of complaints about how the source selection proceeded. Whatever system is used, however, it cannot simply be used to check boxes. A proactive awareness of how the record created by this process will look to outsiders should lead to more discipline in the maintenance of the records required to justify decisions and in the maintenance of only these records.

Tailor Quality Assurance to the Needs of a Less-Experienced Source Selection Workforce

The hiatus in hiring during the 1990s has left the Air Force with a serious shortage of contracting personnel with 10 to 20 years of experience.[20] This is the cohort that has traditionally carried the heaviest load in executing day-to-day acquisition activities of the kind experienced in source selections. Entry-level hiring today will not refill the ranks of this group for another decade or more. Moreover, a significant portion of the cohort with more than 20 years of experience, which the Air Force has traditionally relied on heavily for training young people, is nearing retirement age. Given these demographic facts, what can the Air Force do today to enhance the capabilities of the people responsible for source selection? Our discussions with personnel throughout the Air Force supported the following findings.

The best form of training for a major source selection is real-world experience. However, as the Air Force has conducted fewer complex source selections, opportunities for such training have decreased. To some degree, case-based training and role-playing exercises tailored to the circumstances of a specific new source selection can help substitute

---

[20] Kayes, 2008, Chart 29.

for experience in real source selections.[21] Just-in-time, on-the-job training based on role playing can teach inexperienced personnel specific new skills that they can apply immediately in an upcoming source selection. Role playing also offers personnel an opportunity to see the consequences of errors quickly without imposing large costs on the Air Force when they make such errors during training. In the same way that flight simulators allow pilots to prepare safely for high-risk situations in flight, role playing allows trainees to face serious challenges and learn from them safely. Case-based training and role-playing exercises are time consuming and costly, in terms of both the time of the participants and the cost of the trainers.[22] But if coordinated closely with the terms of an upcoming source selection, they can help prepare new players for roles in major source selections in the future.

Experienced, technically skilled coaches can help back up inexperienced personnel through the course of a source selection by watching over their shoulders as they go through their day-to-day activities. In principle, such coaches could also provide the training described above. Presence on site keeps the coach knowledgeable about the status of the source selection and thus prepared to detect trouble and help remedy it when it occurs. A coach might even participate occasionally in a source selection if events become too demanding.

---

[21] Case-based training uses the particulars of an upcoming source selection to frame instruction. The more closely the instructor can match the issues raised during training to the issues students will have to resolve in the actual source selection, the better. As an extreme case, one might imagine an instructor acting as a coach who continues to work with students through the course of a source selection, helping them address each specific issue as it arises. Role playing during case-based training typically asks all the members of a team to play their expected roles through the course of an exercise. This helps each member learn his or her individual role and, often more important, teaches members where they will sit in the team as the source selection proceeds and how they should interact to succeed.

[22] Of course, it is worth spending a great deal more money on a flight simulator than on just-in-time training. But cost-effective, just-in-time training can be tailored to the characteristics of an upcoming source selection. Presumably, more training would be warranted where the risks to the Air Force in a source selection are higher.

The Right Kind of External Quality Assurance Can Help

As the Air Force goes forward with its plans to use external quality assurance teams more aggressively, the CSAR-X and KC-X programs offer insights into how best to do so. Personnel associated with both programs believe that external oversight is more likely to add value if it displays four basic characteristics.

First, it is timely. Any oversight, internal or external, should occur quickly enough after an action occurs to fix any errors embodied in that action—whether that be preparation of the request for proposal, management of evaluation notices, execution of evaluation, or justification of final decisions. For example, once the requests for proposal were finalized in both source selections, both program offices found themselves committed to conducting evaluations that differed from those they had intended to conduct because this realization came too late.

Second, it is well informed. When external advisors arrive as short-term visitors, they should arrive well informed about the issues in the source selection and the status of the source selection. They have a limited time on site to add value; the personnel working within the source selection have limited time to learn from them while continuing their primary tasks in the source selection. Asking already overworked staff to withdraw from their ongoing source selection responsibilities to educate visitors, who might be somewhat disengaged by their off-site status to begin with, can easily cost more than the value it adds.

Third, it is technical. It is well informed about the technical issues relevant to GAO's oversight of source selection decisions. It is current on recent lessons learned. It can engage junior, less-experienced staff on the working level with detailed support on specific issues. It does not emphasize high-level, gray-beard strategic thinking unless the resulting advice can come early enough in the source selection to affect its overall strategy. GAO emphasizes fine points. The review should mainly respond to that emphasis.

Finally, it is hands-on. It addresses specific decisions and specific documentation in near real time, with the intent of assuring relevant decisions and documentation and training personnel within the source selection to generate similarly concrete decisions and documents after the team leaves.

External quality assurance teams cannot replace on-site expertise. They can offer a different set of eyes with perhaps a slightly different interpretation of what to watch for. They can offer input based on expertise that would otherwise be too costly to commit to a source selection full time. But at the end of the day, even this higher-level support coming from a distance will be most likely to add net value if it is timely, well-informed, technical, and hands-on—just like the coaches on site.

New Data Could Help the Air Force Target Its Efforts

We believe that sophisticated protests will be more likely to occur in large, complex acquisitions with high stakes for the participants, but we cannot be much more specific without more information on when such protests occur. We have no solid information on the costs they are likely to impose on the Air Force. The more the Air Force knows about when they will occur and how they will likely affect the Air Force, the more effectively the Air Force can plan when to introduce changes such as those discussed above and how far to carry them.

A relatively simple way to gather such information is to track future protests sustained by GAO and collect information on their characteristics and the costs that they impose. Sophisticated protests appear to be designed to force the Air Force into a GAO review where a well-prepared protest team expects that it will outperform an Air Force defense well enough to secure a GAO sustainment on at least one issue. So monitoring the relatively small number of sustained protests is a natural place to start collecting data to prepare for future sophisticated protests.

Over time, data collected for this purpose could potentially serve the Air Force in another way. When GAO sustains a protest, it reaches a judgment that the Air Force has made a large enough error potentially to change the outcome of the source selection in question. Is the error in fact large enough to change the outcome? If it is, the error has presumably prejudiced the ultimate winner, who received the contract only because the Air Force corrected its error. If not—if the sustained protest has no effect on the ultimate winner—the Air Force error was probably not serious enough to prejudice the offeror whose protest was sustained.[23] A preliminary Air Force review of recent sustained protests found that many sustained protests do not lead to changes in the winner of an Air Force competition, suggesting that GAO has tended to misjudge the degree of prejudice associated with an Air Force error in these protests.[24]

If more-systematic evidence on how sustainments affect the ultimate winner of Air Force source selections supported this observation, the Air Force could potentially use such evidence, along with evidence on the cost imposed by such unwarranted sustainments, to inform GAO and the Congress of the effects of such sustainments.[25] As a fundamental element of its approach to reviewing bid protests, GAO seeks a "balanced" approach that protects offerors that do experience real prejudice in federal source selections without imposing undue costs on other participants in these source selections when GAO determines that prejudice was not present.[26] The Air Force could use evidence of the kind described here to test how well GAO's approach to balance has worked in practice. If GAO has in fact sustained too many protests

---

[23] We condition these statements, speaking of "presumably prejudiced" and "probably not serious," because, following a protest, the Air Force can never simply correct an error and rerun precisely the same competition it ran earlier. The competition following the protest is necessarily a new competition that takes place in new circumstances. So the result of any second competition cannot tell us with certainty that prejudice was present in the first competition. But the more often the second competition fails to overturn the decision of the first, the more likely it is that material errors are not occurring in the initial competitions. Such a pattern would suggest that GAO balances its decisions too much in favor of protesters.

[24] Kayes, 2008, Charts 15–16.

[25] In principle, to speed development of a meaningful body of evidence, the Air Force could collect more historical data as well as data on sustainments in source selections run by other federal agencies. Either approach is more costly than gathering information on Air Force sustainments as they occur. Gathering reliable historical data well after the fact on the costs imposed by unwarranted sustainments will be particularly challenging.

[26] U.S. Government Accountability Office, *Report to Congress on Bid Protests Involving Defense Procurements*, File B-401197, Washington, D.C., April 9, 2009.

where it turns out that prejudice was not present, the Air Force would have a strong empirical basis for arguing that GAO is not implementing the balance it seeks as effectively as it could.

CHAPTER FOUR

Closing Thoughts

In the wake of the high-visibility sustained protests that resulted from the Druyun scandal[1] and then in the CSAR-X and KC-X programs, it is easy to lose sight of the Air Force's broader experience with GAO bid protests. During the 1990s, the number of unwarranted protests dropped markedly, leaving the Air Force in a position to focus on protests that were more likely to require corrective action. Since 2001, the numbers of corrective actions and of merit reviews per 1,000 contract awards have slowly dropped. The Air Force should be careful to protect the policies and practices that have supported this pattern of steadily improving performance.

The threat manifested in the CSAR-X and KC-X programs appears to be relatively new in character and so does justify significant adjustments in policies and practices in appropriate circumstances.

---

[1] Darlene A. Druyun served as principal deputy assistant secretary of the Air Force for acquisition and management from 1993 until 2003, when she retired and took a position as vice president and deputy general manager of Boeing's missile defense systems. In 2004, she pleaded guilty to "one count of conspiracy for negotiating a job with Boeing [while] overseeing its business with the Pentagon." Subsequently, she admitted that, while in the government, she had favored "Boeing in the selection and pricing of several major projects, including a $20 billion leasing agreement for 100 airborne tankers, a 2002 reworking of a NATO early warning system, a $4 billion upgrading of the C-130 aircraft, and a $412 million payment on a C-17 contract." She did these things to "obtain jobs for herself, her daughter and her son-in-law." When the full scope of her wrong-doing became apparent, DoD responded by reviewing all contracts that she had overseen since 1993. What has come to be known as "the Druyun scandal" was the most serious breach of government acquisition rules in decades. Leslie Wayne, "Ex-Pentagon Official Gets 9 Months for Conspiring to Favor Boeing," *New York Times*, October 2, 2004.

But we expect this threat to present itself in a relatively small number of acquisitions—the large, complex ones, with relatively large stakes for the participants—and it is probably more likely when the participants understand how to pursue sophisticated protests. Such protests will continue and could increase in number until the Air Force demonstrates that it can effectively counter them. The Air Force should focus its countermeasures on the places where the threat is greatest. That should make it easier to tailor the countermeasures to the circumstances and so to choose the set of measures best suited to helping the Air Force avoid future costs and delays, such as those associated with the protests sustained in the CSAR-X and KC-X source selections.

In particular, the Air Force can take the following steps to reduce the risks associated with sophisticated protests:

- Recognize a bid protest as an adversarial proceeding with finely tuned rules. Give greater attention to how GAO views a bid protest. Expect GAO to view the priorities associated with running a source selection differently than the Air Force does. Be prepared to pursue the Air Force's interests within the constraints imposed by GAO's priorities.
- Simplify and clarify selection criteria and priorities. The Air Force is already moving aggressively in this direction. The new approach that WSARA prescribes for capability and requirements determination could help clarify the relative importance of requirements in ways that promote this goal.
- Focus formal cost estimates on the instant contract. Again, the Air Force is already moving aggressively in this direction. As it does so, it should be clear in its source selections how and why the cost estimates it uses to discriminate among proposals differ from the cost estimates it uses in its submissions to the Defense Acquisition Board.
- Tighten discipline throughout the source selection. The Air Force plans to rely more heavily on external review processes to enhance discipline. External review can help; greater involvement of attorneys as part of any external review should be especially helpful. But the issues arising in sophisticated protests can ultimately

be addressed only by hands-on, close attention to detail that an external review team cannot perform. Tools are available to support discipline and simplify internal review. New forms of training and coaching can also help.

- Gather new data to help the Air Force target its efforts. More attention to the costs imposed by different forms of protest could help the Air Force determine how much to spend to avoid these costs. Better data on the extent to which sustained protests actually change who wins a competition could help the Air Force inform GAO about when an error is actually likely to prejudice any offeror and hence justify a sustainment.

Bibliography

Acquisition Advisory Panel, *Report to the Office of Federal Procurement Policy and the United States Congress,* Final Panel Working Draft, Washington, D.C., December 2006.

Bailey, Jeff, and David M. Herszenhorn, "Boeing Says It Will Protest Tanker Deal," *New York Times,* March 11, 2008.

Boockholdt, Kathy, "Source Selection Lessons Learned," briefing, Washington, D.C.: Office of the Air Force Acquisition Center of Excellence (SAF/ACE), December 10, 2008.

Boyd, John R., "Discourse on Winning and Losing," briefing, 1976. As of November 30, 2010:
http://www.ausairpower.net/JRB/intro.pdf

Burton, Robert A., "Analysis of KC-X Tanker Draft RFP for Consistency with the Weapons System Acquisition Reform Act of 2009," white paper, Washington, D.C.: Venable, L.L.P., October 19, 2009.

Butler, Amy, "Air Mobility Command to Begin Search for KC-135 Tanker Replacement," *Inside the Air Force,* December 17, 1999.

———, "Air Force Mulling Replacement for Aging, Maintenance-Needy KC-135," *Inside the Air Force,* May 4, 2001.

Buy American Act, 41 U.S.C. § 10a–10d, 1933.

Cahlink, George, "Ex-Pentagon Procurement Executive Gets Jail Time," *Government Executive,* October 1, 2004.

Camm, Frank, Mary E. Chenoweth, John C. Graser, Thomas Light, Mark A. Lorell, Rena Rudavsky, and Peter Anthony Lewis, *Government Accountability Office Bid Protests in Air Force Source Selections: Evidence and Options,* Santa Monica, Calif.: RAND Corporation, DB-603-AF, 2012. As of January 24, 2012:
http://www.rand.org/pubs/documented_briefings/DB603.html

Christie, Gary E., Dan Davis, and Gene Porter, *Air Force Acquisition: Return to Excellence,* CNA Independent Assessment, CRM D0019891.A2/Final, Alexandria, Va., February 2009.

Congressional Research Service, *The Air Force KC-767 Tanker Lease Proposal: Key Issues for Congress,* Washington, D.C.: The Library of Congress, RL32056, updated September 2, 2003.

"CSAR-X Request Cut by $42 Million, Stipulations Placed on FY06 Funds," *Inside the Air Force,* December 23, 2005.

"CV-22 Likely to Take Over All Current Air Force Rotorcraft Missions," *Inside the Air Force,* November 27, 1998.

Dahmann, Judith S., and Mike Kelley, *Systems Engineering During the Materiel Solution Analysis and Technology Development Phases,* white paper, Washington, D.C.: Office of the Director, Defense Research and Engineering, System Engineering, September 2009.

Defense Information Services Agency (DISA), *Acquisition Deskbook: Source Selection,* Washington, D.C., March 2000.

Donley, Michael B., "Strengthening the Acquisition Process," memorandum, Secretary of the Air Force, Washington, D.C., July 18, 2008a; not available to the general public.

———, "Strengthening the Acquisition Process (Your memo, dated 2 Sep 08)," memorandum, Secretary of the Air Force, Washington, D.C., September 9, 2008b; not available to the general public.

Edwards, Vernon J., *Source Selection Answer Book,* 2nd ed., Vienna, Va.: Management Concepts, 2006.

Federal Acquisition Regulation, Subpart 7.5, "Inherently Governmental Functions." As of November 16, 2009:
https://www.acquisition.gov/far/05-06r1/html/Subpart%207_5.html

———, Subpart 14.2, "Solicitation of Bids." As of May 18, 2011:
http://farsite.hill.af.mil/reghtml/regs/far2afmcfars/fardfars/far/14.htm.

Gates, Dominic, "Pitching 777 for Tanker Contest Could Mean More Jobs at Boeing Everett Plant," *Seattle Times,* June 17, 2009.

Golden, Michael R., "Evaluating Cost or Price in Competitions: Challenges Abound," Washington, D.C.: U.S. Government Accountability Office, paper presented at DoD Cost Analysis Symposium 42, Williamsburg, Va., February 2009.

Grunbaum, Rami, "Opponents Call Boeing's Plane 'Frankentanker,'" *Seattle Times,* January 27, 2008.

Hawthorne, Skip, "Risk Based Source Selection," briefing at PEO/SYSCOM Conference, Washington, D.C.: Office of the Under Secretary of Defense for Acquisition, Technology and Logistics, Defense Procurement Acquisition Policy, November 7, 2006.

Ilg, Scott, "Best Value Source Selection Trade-Offs," briefing, Kaiserslautern, Germany: Defense Acquisition University, 2006.

Kadish, Ronald (Lt Gen, USAF, Ret.), *Defense Acquisition Performance Assessment: Report by the Assessment Panel of the Defense Acquisition Performance Assessment Project for the Deputy Secretary of Defense*, Washington, D.C., January 2006.

Kayes, Brett N. (Capt, USAF), "Air Force GAO Protest Trend Analysis," briefing chart, Washington, D.C.: Deputy Assistant Secretary (Contracting) (Contracting Operations Division) (SAF/AQCK), updated September 19, 2008.

Kornreich, Doug, "Bid Protests: How to Avoid Them, and How to Win Them," briefing, Washington, D.C.: Department of Health and Human Services, Office of General Counsel, General Law Division, 2004.

Kreisher, Otto, "Is CSAR Really Nothing 'Special'?" *Air Force Magazine*, November 2009.

Leung, Rebecca, "Cashing In For Profit? Who Cost Taxpayers Billions In Biggest Pentagon Scandal In Years?" CBS News, January 5, 2005. As of February 8, 2010: http://www.cbsnews.com/stories/2005/01/04/60II/main664652.shtml

Liang, John, "Wall Street Analysts Surprised by Lockheed Team's Presidential Helicopter Win," *DefenseAlert-Daily News*, January 31, 2005.

Light, Thomas, Frank Camm, Mary E. Chenoweth, Peter Anthony Lewis, and Rena Rudavsky, *Analysis of Government Accountability Office Bid Protests in Air Force Source Selections over the Past Two Decades*, Santa Monica, Calif.: RAND Corporation, TR-883-AF, 2012. As of January 24, 2012: http://www.rand.org/pubs/technical_reports/TR883.html

Manuel, Kate M., and Moshe Schwartz, *GAO Bid Protests: An Overview of Timeframes and Procedures*, Washington, D.C.: Congressional Research Service, February 11, 2009.

Martin, Edward C., "Source Selection Improvements," briefing, Wright-Patterson AFB, Ohio: Aeronautical Systems Center Acquisition Center of Excellence, April 10, 2007.

Martin, Edward C., and Daniel C. Fulmer, "What's New in Air Force Source Selection," Aeronautical Systems Center Acquisition Center of Excellence, briefing, World Congress, National Contract Management Association, Session #711, Cincinnati, Ohio, April 13–16, 2008.

Matishak, Martin, "Pentagon Appeals Host of CSAR-X Changes Proposed by Lawmakers," *Inside the Air Force*, October 28, 2005.

————, "Lawmakers Want Detailed Cost, Schedule Plans: CSAR-X Request Cut by $42 Million, Stipulations Paced on FY06 Funding," *Inside the Air Force,* December 23, 2006.

McCain, John (Senator), Letter to the Honorable Robert M. Gates, U.S. Senate Committee on Armed Services, Washington, D.C., October 29, 2009.

"New Helicopters Possibly in Store for Enhanced Combat Search and Rescue," *Inside the Air Force,* February 12, 1999.

Payton, Sue C., "Implementation Plan for Strengthening the Acquisition Process, Memorandum," Washington, D.C.: SAF/AQ, September 15, 2008a; not available to the general public.

————, "Strengthening the Acquisition Process (SECAF Memorandum, July 18, 2008)," memorandum with attachments, "Point Paper on KC-X Source Selection: Lessons Learned," "Air Force Acquisition Quick Look: Terms of Reference," Washington, D.C.: SAF/AQ, September 2, 2008b; not available to the general public.

————, "Strengthening the Acquisition Process (Draft Recommendations), Memorandum," Washington, D.C.: SAF/AQ, September 15, 2008c; not available to the general public.

Pfleger, Katherine, "Lawmakers Consider Air Force as Boeing Commercial-Plane Customer," *Associated Press Newswires,* September 30, 2001, quoted in Congressional Research Service, September 2, 2003.

Pollock, Rob, "Weekly Assessment of Implementation Plan Progress to Strengthen the Acquisition Process," briefing, Washington, D.C.: Acquisition Chief Process Office (SAF/ACPO), December 2, 2008.

Schanz, Marc V., "Gates Hits Reset Button on CSAR-X," *Air Force Magazine,* April 7, 2009.

Schwartz, Moshe, and Kate M. Manuel, *GAO Bid Protests: Trends, Analysis, and Options for Congress,* Washington, D.C.: Congressional Research Service, February 2, 2009.

Scott, Bethany, "Air Force Eyes Replacement for Aging Pave Hawk Helos," *National Defense Magazine,* September 2001.

Slate, Alexander R., "Best Value Source Selection: The Air Force Approach," *Defense AT&L,* September–October 2004, pp. 52–55.

Tiron, Roxana, "GAO Rules in Favor of CSAR-X Protesters," *The Hill,* August 30, 2007.

Tirpak, John A., "Tanker Twilight Zone," *Air Force Magazine,* February 2004.

U.S. Air Force, Request for Proposal for the Combat Search and Rescue Recovery Vehicle (CSAR-X) System Development and Demonstration (SDD), CSAR-X Low Rate Initial Production (LRIP) and Initial Production Options, Solicitation FA8629-06-R-2350, Wright-Patterson AFB, Ohio: Aeronautical Systems Center, 2005.

————, Request for Proposal for the KC-X Tanker Replacement Program, Solicitation FA8625-07-R-6470, Wright-Patterson AFB, Ohio: Aeronautical Systems Center, January 29, 2007.

————, "USAF Annual Bid Protest Update, FY09Q1," briefing, Washington, D.C.: SAF/AQC, January 1, 2009a.

————, Source Selection, Mandatory Procedure MP5315.3, Washington, D.C.: SAF/AQCP, March 2009b.

————, Acquisition Improvement Plan, Washington, D.C.: SAF/AQ, May 4, 2009c.

U.S. Air Force Acquisition Chief Process Office, United States Air Force "Strengthening the Acquisition Process" Final Report, Report to the Secretary of the Air Force on the "Strengthening the Acquisition Process," Washington, D.C., 2009; not available to the general public.

U.S. Army, Army Source Selection Manual, Addendum AA to AFARS, May 16, 2008.

U.S. Department of Defense, Addendum to the Defense Acquisition Structures and Capabilities Review, Pursuant to Section 814, National Defense Authorization Act, Fiscal Year 2006, Washington, D.C., June 2007a.

————, Report of the Defense Acquisition Structures and Capabilities Review, Pursuant to Section 814, National Defense Authorization Act, Fiscal Year 2006, Washington, D.C., June 2007b.

————, "Executive Report," Understanding the Problem Subcommittee, Source Selection Joint Analysis Team, Washington, D.C., April 24, 2009a.

————, "Guidance Joint Framework," briefing, Source Selection Joint Analysis Team, Washington, D.C., April 24, 2009b.

————, "Recommendations for Best Practices," Source Selection Joint Analysis Team, Washington, D.C., April 24, 2009c.

————, "Source Selection Policy and Guidance Assessment—Initial Report: Executive Summary," Source Selection Joint Analysis Team, Guidance Subcommittee, Washington, D.C., April 24, 2009d.

————, "Understanding the Problem Subcommittee Briefing," Source Selection Joint Analysis Team, Washington, D.C., April 24, 2009e.

U.S. Department of Defense, Inspector General, *Acquisition of the Boeing KC-767A Tanker Aircraft,* D-2004-064, Washington, D.C., March 29, 2004.

———, *Air Force KC-X Aerial Refueling Tanker Aircraft Program,* D-2007-103, Washington, D.C., May 30, 2007.

U.S. General Accounting Office, *U.S. Combat Air Power: Aging Refueling Aircraft Are Costly to Maintain and Operate,* GAO/NSIAD-96-160, Washington, D.C., August 1996.

———, *Military Aircraft: Observations on the Proposed Lease of Aerial Refueling Aircraft by the Air Force,* GAO-03-923T, Washington, D.C., September 4, 2003.

———, *DoD Needs to Determine Its Aerial Refueling Aircraft Requirements,* GAO-04-349, Washington, D.C., June 2004.

U.S. General Services Administration, Department of Defense, and National Aeronautics and Space Administration, *Federal Acquisition Regulation,* Washington, D.C., March 2005.

U.S. Government Accountability Office, "Bid Protest Regulations," website, n.d. As of February 8, 2010:
http://www.gao.gov/decisions/bidpro/bid/bibreg.html

———, *Bid Protests at GAO: A Descriptive Guide,* 8th ed., GAO 06-797SP, Washington, D.C., 2006.

———, Decision, Matter of Sikorsky Aircraft Company, Lockheed Martin Systems Integration-Owego, Files B-299145, B-299145.2, B-299145.3, Washington, D.C., February 26, 2007a.

———, Decision, Matter of Sikorsky Aircraft Company, Lockheed Martin Systems Integration-Owego, Request for Reconsideration, File B-299145.4, Washington, D.C., March 29, 2007b.

———, *Cost Assessment Guide: Best Practices for Estimating and Managing Program Costs,* GAO-07-1134SP, Washington, D.C., July 2007c.

———, *Selected Recent GAO Bid Protest Decisions,* Washington, D.C., updated August 2007d.

———, Decision, Matter of Lockheed Martin Systems Integration-Owego, Sikorsky Aircraft Company, Files B-299145.5, B-299145.6, Washington, D.C., August 30, 2007e.

———, Decision, Matter of the Boeing Company, Files B-311344, B-311344.3, B-311344.4, B-311344.6, B-311344.7, B-311344.8, B-311344.10, B-311344.11, Washington, D.C., June 18, 2008.

————, Decision, Matter of Delex Systems, Inc., File B-400403, October 8, 2008. As of July 26, 2010:
http://www.gao.gov/decisions/bidpro/400403.htm

————, *Report to Congress on Bid Protests Involving Defense Procurements,* File B-401197, Washington, D.C., April 9, 2009.

Van Buren, David M., "Acquisition Improvement Plan Implementation," memorandum, SAF/AQ, Washington, D.C., May 8, 2009.

Warwick, Graham, "Why Boeing's HH-47 Chinook Won the CSAR-X Competition," *Flight,* Flightglobal.com, October 11, 2006.

Wayne, Leslie, "Ex-Pentagon Official Gets 9 Months for Conspiring to Favor Boeing," *New York Times,* October 2, 2004. As of December 1, 2010:
http://www.nytimes.com/2004/10/02/business/02boeing.html?ref=darleen_a_druyun

————, "U.S.-Europe Team Beats Out Boeing on Big Contract," *New York Times,* March 1, 2008.

Weapon System Acquisition Reform Act, Public Law 111-23, May 22, 2009.

Weisgerber, Marcus, "Gates Reopens Tanker Competition, Shifts Authority from Air Force," *Inside the Pentagon,* August 10, 2009.